Calculus Test Forms

Instructions

Tests are an important component of the Saxon program. We believe that concepts and skills should be continually tested. However, tests should only be administered after the concepts and skills have been thoroughly practiced. Therefore, we recommend that tests be administered according to the testing schedule which is printed on the back side of this page.

Note: Due to the amount of scratch work necessary to solve some calculus problems, it may be necessary for a student to show his/her work on scratch paper. If tests are to be reused in the future, the student should write his/her answers on a separate sheet of paper.

Calculus

Testing Schedule

Test to be administered:	Covers material up through:	Give after teaching:
Test 1	Lesson 2	Lesson 6
Test 2	Lesson 6	Lesson 10
Test 3	Lesson 10	Lesson 14
Test 4	Lesson 14	Lesson 18
Test 5	Lesson 18	Lesson 22
Test 6	Lesson 22	Lesson 26
Test 7	Lesson 26	Lesson 30
Test 8	Lesson 30	Lesson 34
Test 9	Lesson 34	Lesson 38
Test 10	Lesson 38	Lesson 42
Test 11	Lesson 42	Lesson 46
Test 12	Lesson 46	Lesson 50
Test 13	Lesson 50	Lesson 54
Test 14	Lesson 54	Lesson 58
Test 15	Lesson 58	Lesson 62
Test 16	Lesson 62	Lesson 66
Test 17	Lesson 66	Lesson 70
Test 18	Lesson 70	Lesson 74
Test 19	Lesson 74	Lesson 78
Test 20	Lesson 78	Lesson 82
Test 21	Lesson 82	Lesson 86
Test 22	Lesson 86	Lesson 90
Test 23	Lesson 90	Lesson 94
Test 24	Lesson 94	Lesson 98
Test 25	Lesson 98	Lesson 102
Test 26	Lesson 102	Lesson 106
Test 27	Lesson 106	Lesson 110
Test 28	Lesson 110	Lesson 114
Test 29	Lesson 114	Lesson 114

1. The wheel on Michelle's unicycle revolved 30 times every minute as she rode down the straight path. If the radius of the wheel of her unicycle was 50 centimeters, how fast was Michelle's unicycle traveling along the path in centimeters per minute?

2. The wagon was pulled down the road at a speed of 50 feet per minute. If the radius of each of the wheels of the wagon was 1 foot, how many revolutions were each of the wheels making per minute?

Evaluate:

3. $\sin^2 \frac{\pi}{3} + \tan \frac{\pi}{4}$

4. $\csc 30° - \sec^2 \frac{\pi}{6}$

Simplify:

5. $(\cos^2 \theta)(\sec \theta)(\tan \theta)$

6. $\frac{\cot \theta \cos \theta}{\csc \theta}$

7. Write the contrapositive of the statement: "If there are clouds in the sky, then it is raining."

8. Is the following argument valid? Explain your answer.
If there are clouds in the sky, then it is raining.
It is not raining.
Thus, there are no clouds in the sky.

9. Write the point-slope form of the equation of the line which passes through (3, 1) and is parallel to the line $3y = 2x - 3$.

10. Multiply $2x^2 - 3x + 1$ by $3x + 1$ and then simplify.

11. Find the values of x which satisfy the equation $x^2 + x = 1$.

Simplify:

12. $\dfrac{1}{1 + \dfrac{1}{1 + \dfrac{1}{2}}}$

13. $3\sqrt{\frac{1}{3}} - 3\sqrt{3} + \sqrt{27}$

14. Write $\frac{2i - 4}{1 + i}$ as a fraction whose denominator is a rational number.

15. Evaluate: $\sum_{i=-1}^{1} (1 - 2^i)$

Concept Review:

Find x in the following drawings:

16. $m\widehat{AB} = (5x + 50)°$
$m\widehat{BC} = (3x + 30)°$
$m\angle BPA = (2x)°$

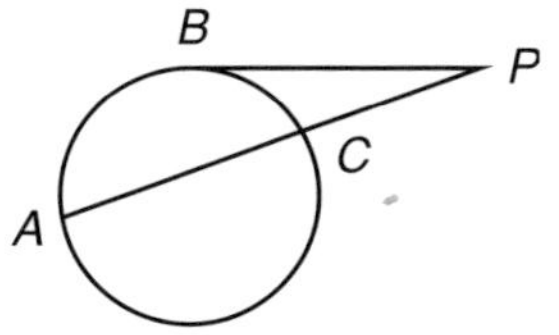

17.

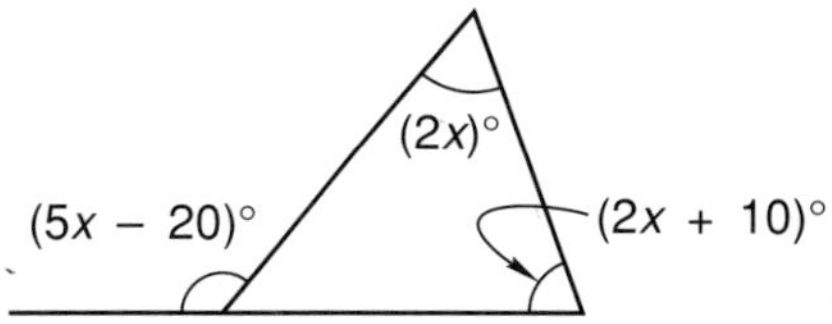

1. Two pipes are used to fill a large tank. When used alone, pipes A and B take 4 hours and 6 hours, respectively, to fill the tank. How many hours would it take to fill the tank if the two pipes were used simultaneously?

2. Nine years from now, Judy will be three times as old as she was 11 years ago. How old is she now?

3. If a wheel rotates at a rate of 4 revolutions per second, find the number of radians the wheel rotates every minute.

4. Solve for x in terms of L:

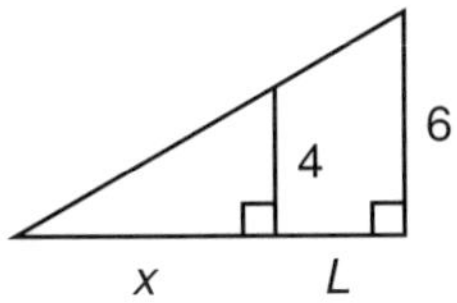

5. Solve for a, b, and c.

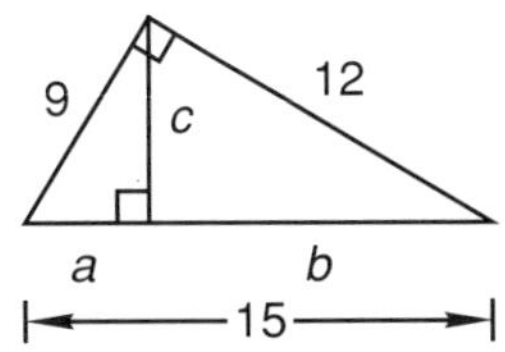

6. Shown is a unit circle centered at the origin of the coordinate plane. Find the coordinates of p.

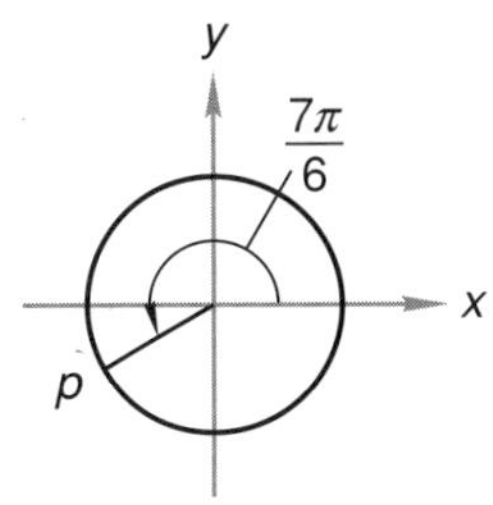

7. Write the equation of the sinusoid shown in terms of the sine function.

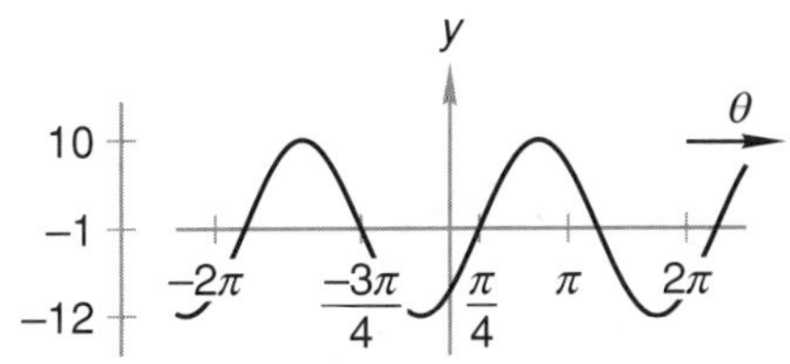

8. Shown is a function machine f where only a few input values with their corresponding output values are given.

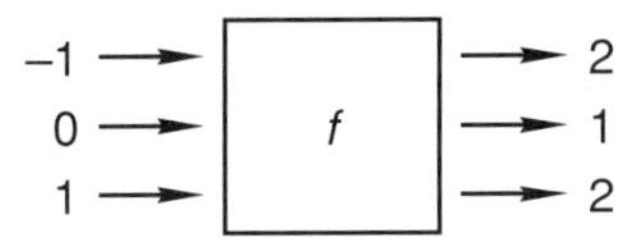

Which of the following could be the equation of f?

a. $f(x) = -2x$ b. $f(x) = -x + 1$

c. $f(x) = 2^{-x}$ d. $f(x) = x^2 + 1$

9. Find the domain and range of $y = 2 + \sqrt{x}$.

10. Simplify: $\dfrac{\cos(-x)\sec(-x)}{\cot(-x)}$

11. Evaluate: $\sin\left(-\dfrac{\pi}{4}\right) + 2\tan^2\left(\dfrac{\pi}{6}\right)$

12. Divide $x^5 + x^2 + 1$ by $x^3 + 1$.

13. Find the values of x for which $2x^2 - 3x - 1$ is zero.

Concept Review:

14. If x is a positive real number, compare: A. 10% of x B. $(0.10)x$

15. Solve for y: $AE = y$
$EB = y + 2$
$DE = 2y$
$CE = y - 1$

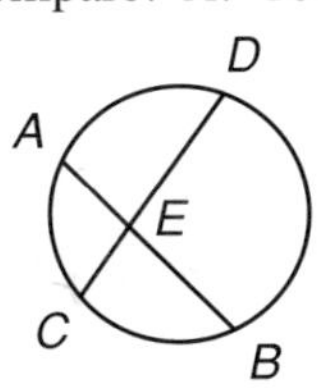

Test 3, Lesson 10, Form A **SHOW YOUR WORK** **Name:** ____________________

1. Christy drove at 50 miles per hour for h hours but arrived g hours late. How fast should Christy have driven to have arrived on time?

2. If Chad had worked alone, he could have filled the hole with dirt in k hours, and if Brian had worked alone, he could have filled the hole in 5 hours. If they had chosen to work together, how long would it have taken both to fill h holes with dirt?

3. The number of people listening attentively varied directly with the interest index. If P people listened attentively when the interest index was I, how many people would have been listening attentively if the interest index had been J?

Sketch the graphs of the following equations:

4. $y = e^{-x}$

5. $y = -e^{-x}$

Simplify the following expressions:

6. $\cos\left(\frac{\pi}{2} - \theta\right)\sec\theta$

7. $\dfrac{\tan^2 x + 1}{\sec^2(-x)}$

8. Find the equation of the quadratic function f whose x intercepts are 1 and -2 and whose y intercept is -2.

9. Sketch the graph of $y = -(x - 2)^2 + 1$.

10. Given are two triangles whose angles and sides are as shown. Express a in terms of x, y, and b.

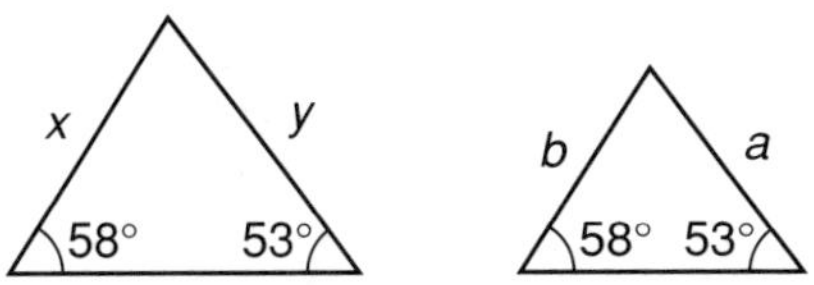

11. Write the equation of the sinusoid shown in terms of the cosine function.

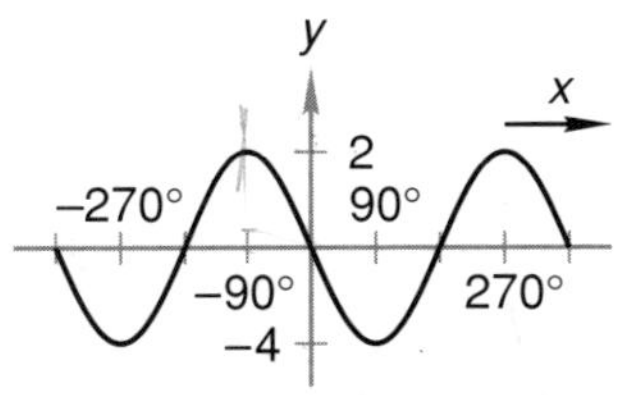

12. Shown is a machine h which maps various input numbers to their corresponding output numbers. Determine whether or not the following machine is a function machine. Justify your answer.

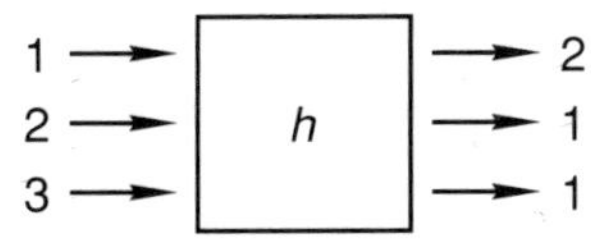

13. Evaluate: $\sum_{i=-1}^{1}\left(\frac{1}{2}\right)^{-i}$

14. Given: $f(x) = 2x$, evaluate and simplify: $\dfrac{f(x + h) - f(x)}{h}$

Concept Review:

15. Find x in the figure shown where O is the center of the circle shown and $m\widehat{AB} = 100°$.

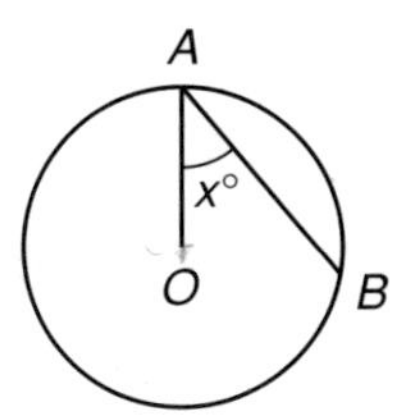

1. The length of a rectangle is three times its width. If the area of the rectangle is 75 square centimeters, find the length and width of the rectangle.

2. On the assembly line, W workers worked H hours to produce P items. If Y additional workers work on the assembly line, how long would it take to produce $P + 4$ items?

A calculator may be used for problems 3–5:

3. Find x where $10^x = 4$.

4. Find x where $e^x = 5$.

5. Solve $\log_2 8 = 3b + 1$ for b.

6. Graph f where $f(x) = [x]$ and evaluate $f(-1.5)$.

7. Graph: $y = |x^2 + x - 2|$.

8. Write the equation of this sinusoid as a sine function.

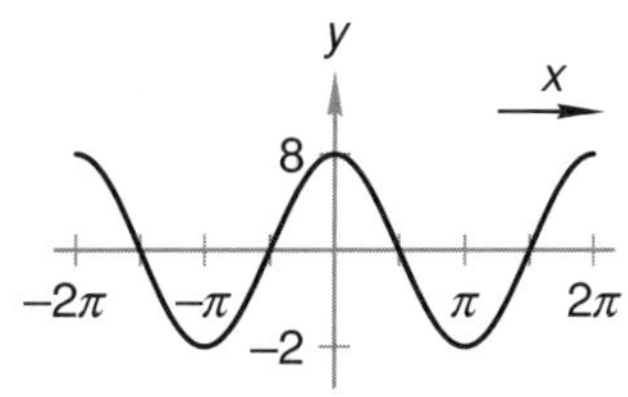

9. Describe the circle whose equation is $x^2 + y^2 + 2y = 0$.

10. Describe the set of all real values of x such that $|x - 2| < 0.05$.

11. Express $-2\angle 120°$ in rectangular form.

12. Graph $y = 2^x$ and $y = 2^{-x}$ on the same coordinate plane.

13. Simplify: $\cot(-x)(\cos x)(\tan^2 x + 1)$

14. Find x in terms of y for the figure shown.

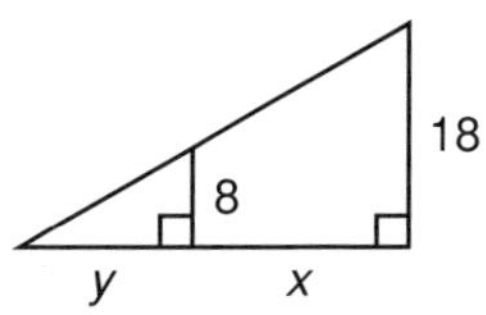

15. Determine whether or not the mapping f: $x \rightarrow |x|$ is a function of x and explain your answer.

Concept Review:

16. Suppose AP and PB are tangents drawn to circle O. Solve for x:
$m\angle APB = x°$
$m\widehat{ADB} = 270°$

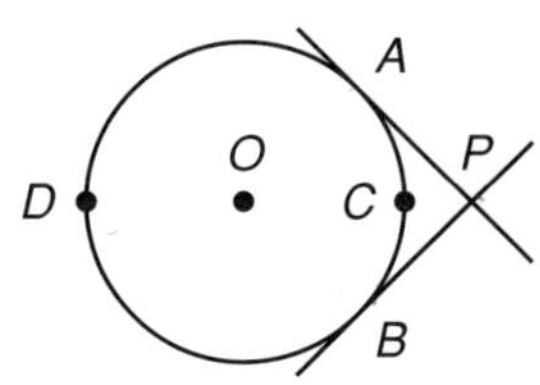

1. Two boats leave the marker at the same time. One of the boats travels northward at a rate of $5N$ miles per hour and the other boat travels westward at a rate of $12N$ miles per hour. What is the distance between the boats 1 hour after both boats have left from the marker?

2. The number of topologists present varied inversely as the square of the number of algebraists present. If 20 topologists were present when 6 algebraists were present, how many topologists would be present if there were 12 algebraists present?

3. Shown is the graph of a function f:

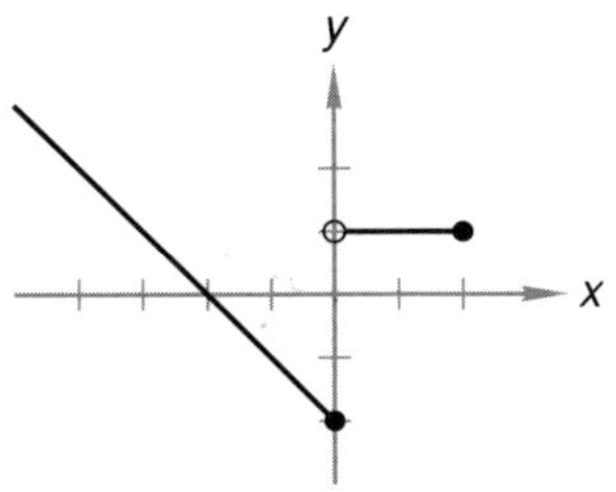

Evaluate the following limits:

a. $\lim_{x\to 0^+} f$ **b.** $\lim_{x\to 0^-} f$

4. Graph $y = [x] - 1$.

5. Suppose $f(x) = x^5 + 2x^4 - 3x^3 + 2x^2 - x + 1$; evaluate $f(2)$ and $f(-1)$ by using synthetic division.

6. Use the rational root theorem to list all the possible rational roots of $f(x) = x^4 + 3x^3 - 3x^2 + 3x - 4$.

7. Use the rational root theorem and synthetic division to find one root of f as defined in problem 6.

8. Sketch the graphs of $y = 3^x$ and $y = \log_3 x$ on the same coordinate plane.

9. Sketch the graphs of $y = \log_2 x$ and $y = -\log_2 x$ on the same coordinate plane.

10. Determine the equation of the centerline and the values of the amplitude, period, and phase angle of the sinusoid whose equation is: $y = 2 + 3 \sin (3x + 45°)$

11. Solve for x: $\log_2 (2x + 1) = 3$

12. Write the standard form of the equation of a circle whose center is $(-1, 2)$ and whose radius is 4.

13. Suppose $\sin A = \frac{1}{2}$ and $\cos A = \frac{\sqrt{3}}{2}$; use the double angle identities to determine the values of $\sin 2A$ and $\cos 2A$.

14. Find the domain and range of $y = \sqrt{x}$.

Concept Review:

15. Solve for t in terms of x, y, and z:

$$AE = t$$
$$EB = x$$
$$CE = y$$
$$ED = z$$

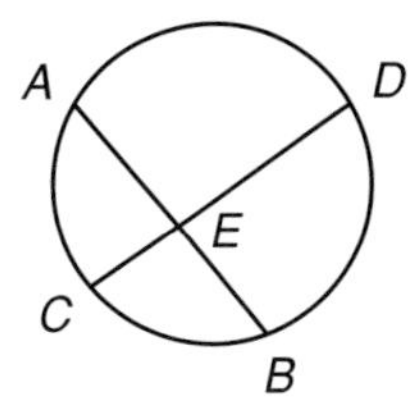

1. The master painter can paint the entire house alone in just D days, but his assistant would require twice as long to paint the house alone. If both worked together, how long would it take to paint k houses?

2. The sum of two numbers is 50. If L denotes the larger of the two numbers, express the product of the two numbers entirely in terms of L.

3. Shown is the graph of a function f.

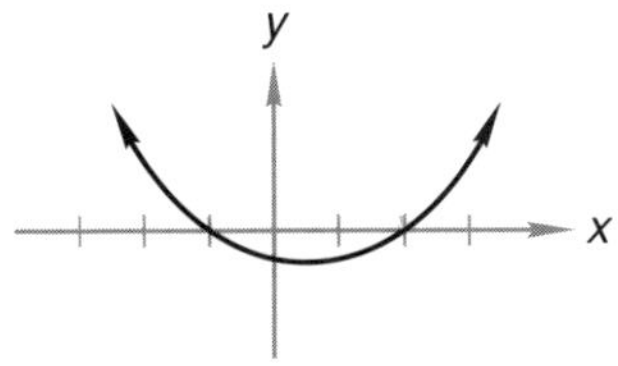

Use interval notation to describe those intervals on which f is positive.

4. If $f(x) = x(x - 1)(x + 2)$, show on a number line those values of x for which f is equal to 0, those intervals on which f is positive, and those intervals on which f is negative.

Solve for x:

5. $\sin^2 x = 1, 0 \leq x < 2\pi$

6. $\sin^2 x + 2 \sin x + 1 = 0, 0 \leq x < 2\pi$

7. Write the following equation of a parabola in the form $(y - k)^2 = 4p(x - h)$: $y^2 - 2y - 4x + 1 = 0$.

8. Find the coordinates of the vertex of the parabola defined in problem 7. Find the value of p. Use the value of p to find the coordinates of the focus and the equation of the directrix.

Evaluate the following limits:

9. $\lim_{t \to 2} \frac{t^2 - 4}{t - 2}$

10. $\lim_{t \to 2} \frac{t^2 - 4}{t + 2}$

11. For which value of k does $y = \cos kx$ have a period of 2?

12. If A is an angle such that $\sin A = \frac{1}{2}$, determine the value of $\sin A \cos \left(\frac{\pi}{2} - A\right)$.

Concept Review:

13. Find the sum of the interior angles of any quadrilateral. That is, find the sum of x, y, z, and t in the figure shown.

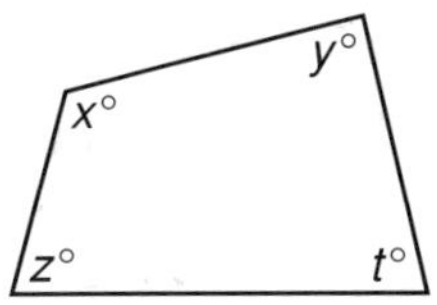

1. Farmer Harvey wants to make a rectangular pasture and wants to use an existing stone wall as one side of the pasture. Fencing is required for the remaining three sides of the pasture. If the pasture has an area of 10,000 square feet and the length of the fence parallel to the wall is L, find the total length of fencing required in terms of L.

2. Find the area of a rectangular field in terms of W if its width is W and its perimeter is 100 yards.

Evaluate the following limits:

3. $\lim_{x\to\infty} \dfrac{2x^3 - x}{1 - 3x^3}$

4. $\lim_{x\to a} \dfrac{x^2 - a^2}{x - a}$

For problems 5–7, f is a function whose graph is shown here:

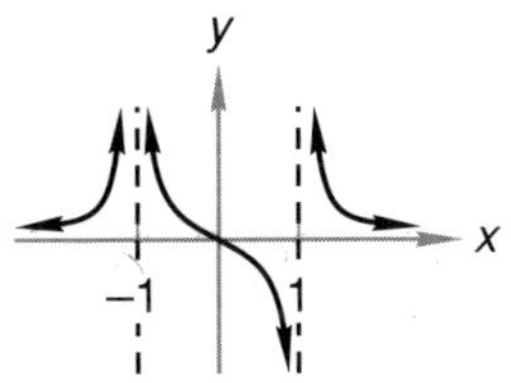

Evaluate:

5. $\lim_{x\to 1^+} f$

6. $\lim_{x\to -1^+} f$

7. Use interval notation to describe those intervals on which f is increasing.

8. Write the standard form of the ellipse whose foci are (4, 0) and (−4, 0) and whose vertices are (5, 0) and (−5, 0).

9. Describe fully and sketch the graph of the ellipse whose equation is: $\dfrac{(x-2)^2}{4} + \dfrac{(y-3)^2}{9} = 1$

For problems 10 and 11, assume that $f(x) = x^3$ and $g(x) = 1 + x$.

10. Write the equation of $\left(\dfrac{f}{g}\right)(x)$ and evaluate $\left(\dfrac{f}{g}\right)(1)$.

11. Describe the domain of $\left(\dfrac{f}{g}\right)$.

In problems 12 and 13, solve for x in terms of y.

12. $y = \ln(x - 1)$

13. $y = e^{x-1}$

14. Use interval notation to describe those intervals on which $f > 0$ where $f(x) = (x - 1)(x + 1)(x + 3)$.

Concept Review:

15. Find the length of the side of a square which can be inscribed in a circle of radius 2.

Test 8, Lesson 30, Form A **SHOW YOUR WORK** **Name:** ____________________

1. On the outbound leg of K miles, young John walked at a brisk pace. On the way back, young John had to run and so doubled his pace. The entire trip took M hours. Express his rate on the outbound leg in terms of K and M.

2. Write in standard form this equation of a hyperbola: $9x^2 - 4y^2 - 18x - 8y - 31 = 0$.

3. Determine the coordinates of the center, the coordinates of the vertices, and the equations of the asymptotes of the hyperbola in problem 2. Graph the hyperbola.

4. Graph the function $f(x) = \frac{1}{x}$. Then, sketch f shifted 2 units down and 1 unit to the right. Write the equation of the new graph.

5. Use a calculator to approximate the value of $\log_6 10$.

6. Find the values of x for which $\log_3 2x > 1$.

7. If $f(x) = x^2$, use the definition of the derivative to find $\frac{d}{dx}f(x)$.

8. If $f(x) = e^x$ and $g(x) = x^2$, write the equation of $(f \circ g)(x)$.

9. Determine the domain and range of f, g, and $f \circ g$ in problem 8.

Evaluate the following limits (use the symbols $+\infty$ and $-\infty$ when their meanings are entirely unambiguous).

10. $\lim_{x \to -2} \frac{1}{2 - x}$

11. $\lim_{x \to 2} \frac{x^2 - 4}{x - 2}$

Solve the following equation for x:

12. $4^{3x-1} = 8^{x+1}$

13. Find the distance between the focus and vertex of a parabola whose equation is $y^2 = 8x$.

14. Simplify: $\dfrac{\sin\theta \cos\left(\frac{\pi}{2} - \theta\right)}{1 - \cos^2\theta}$

Concept Review:

15. If x is a real number, compare: $A.\ x$ $B.\ x^2$

1. The pressure of an ideal gas varies directly with the temperature and inversely with the volume. If the initial pressure, volume, and temperature were P newtons per square meter, L liters, and T Kelvin, what would the pressure be if the volume and temperature were 5 liters and 100 Kelvin, respectively?

2. Express $\sin(A + B)$ and $\cos(A + B)$ as functions of the sine and cosine of A and B.

3. Use problem 2 to develop expressions for $\tan(A + B)$ in terms $\tan A$ and $\tan B$.

4. Find the volume of a sphere whose surface area is 16π m^2.

5. Find the volume of a right circular cone whose base has area 4π cm^2 and whose height is 5 cm.

Use the power rule of differentiation to work problems 6–8.

6. Find $\frac{dy}{dx}$ if $y = \frac{1}{x^2}$

7. Find $f'(x)$ if $f(x) = \sqrt{x}$

8. Find $\frac{ds}{dt}$ if $s(t) = \frac{1}{\sqrt{t^3}}$

9. Find $D_x y$ if $y = x^{13}$

10. Express the four fourth roots of $\frac{1}{2} + \frac{\sqrt{3}}{2}i$ in polar form.

11. Find all the values of x which lie between 0 and 2π which satisfy the equation $\sin 3x = \frac{1}{2}$.

12. The general equation of a conic section is $4x^2 - y^2 - 8x - 2y - 1 = 0$. Write the equation of this conic section in standard form and describe this conic section.

13. Find the coefficient of x^4y in the expansion of $(x + 2y)^5$.

14. Find all those values of x for which $\log_3(x + 2) > 1$.

Concept Review:

15. If $m\widehat{AC} = 60°$, what is the measure of angle ADC?

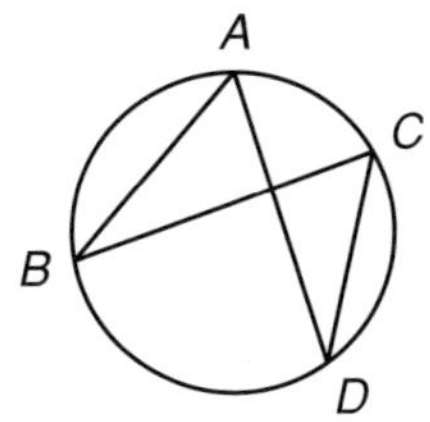

Test 10, Lesson 38, Form A **SHOW YOUR WORK** **Name:** ____________________

1. The volume of the glob increased exponentially. The glob occupied 100 cubic meters at the instant of the big bang, and after 20 minutes, the glob occupied 130 cubic meters. How many cubic meters would the glob occupy 1 hour after the big bang?

Differentiate the functions of problems 2 and 3:

2. $y = \frac{1}{\sqrt{x}} + \frac{1}{3}e^x - 3\cos x$

3. $y = \frac{1}{x} + 3\ln|x| - 2\sin x$

Differentiate as many times as necessary to solve problems 4 and 5:

4. Find $f''(t)$ if $f(t) = 2\sin t - \ln t$

5. Find $\frac{d^3y}{dx^3}$ if $y = 4e^x - 3x^2$

6. Find the equation of the tangent line which can be drawn to the graph of $y = 2x^2$ at the point (1, 2).

7. Begin with the key identities and develop an identity for $\tan(A + B)$.

8. Express the 4 fourth roots of –1 in polar form.

9. Find all the values of x, where $0 \le x < 360°$, which satisfy $\sin 3x + 1 = 0$.

10. Identify and describe the important features of the conic section whose equation is $9x^2 - 4y^2 - 36 = 0$.

11. Describe the set of values of x for which $\log_2(x + 2) > 2$.

12. Use your calculator to approximate the value of $\log_3 5$ to two decimal places. Check your answer on your calculator by raising 3 to your answer.

13. Use interval notation to indicate the intervals on which the graph of f lies below the x axis where $f(x) = x(x + 1)(x + 3)$.

14. Use the rational root theorem to find all the roots of $f(x) = x^3 + 2x^2 - x - 2$.

15. Evaluate: $\lim_{x \to 2} \frac{x^2 + x - 6}{x^2 - 4}$

Concept Review:

16. Solve for x in terms of a and c in the figure shown.

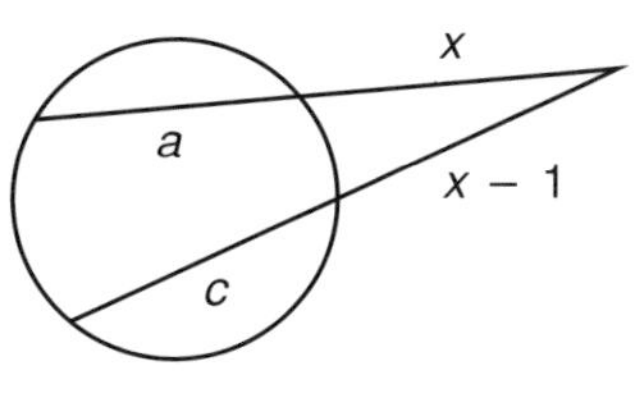

1. Equal-sized squares are cut from the corners of an 8-inch by 8-inch sheet. The resulting flaps are folded up to form an open box. If the length of a side of a square is x, find the volume of the box in terms of x.

2. The speed of the unicorn increased exponentially. At noon, its speed was 20 furlongs per hour, and 2 hours later its speed was 50 furlongs per hour. What would be the speed at 6:00 p.m.?

Use the product rule as required in problems 3 and 4:

3. Find y' if $y = x^2e^x$

4. Find $\frac{dy}{dt}$ if $y = -t \sin t$

Differentiate as many times as necessary in problems 5 and 6:

5. If $s(x) = \frac{1}{\sqrt{x}} + 2 \sin x$, find $s''(x)$.

6. If $f(x) = \ln |x| - 2e^x$, find $f''(x)$.

7. If $f(x) = x^2 - 4$ and $g(x) = \frac{1}{f(x)}$, graph f and g on the same coordinate axes.

Sketch the following functions:

8. $y = \cot x$

9. $y = \frac{x}{(x + 2)(3 - x)}$

10. $y = \sec x$

11. Find the equation of the tangent line which can be drawn to the graph of $y = \sqrt[3]{x}$ at $x = 8$.

12. Find all the values of x which lie between 0 and 2π for which $\sin 2x = \frac{\sqrt{2}}{2}$.

13. Identify and describe the important features of the conic section whose equation is $\frac{(x - 1)^2}{4} + \frac{(y + 2)^2}{9} = 1$.

14. If $f(x) = |x|$, then graph the equation $y = f(x + 1)$.

15. Solve for x: $\ln x - \ln (x + 1) = \ln 2$

Concept Review:

16. Suppose PA is tangent to the circle shown at A and PD is a secant. Solve for x in terms of a and b.
$PA = a$
$PC = b$
$DC = x$

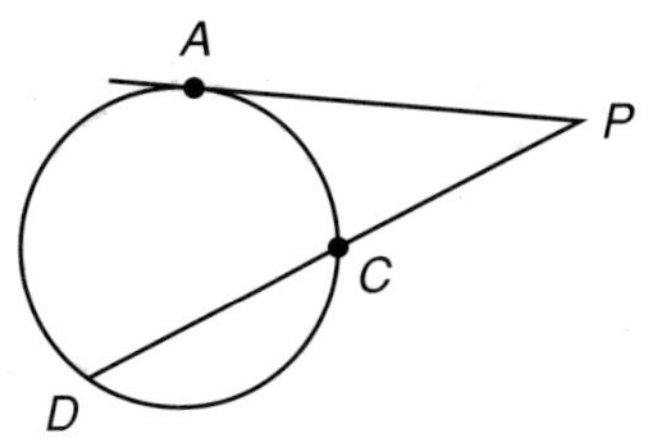

1. The volume of a rectangular solid is 100 cm^3. If the length L and width W of the solid are equal, express the surface area of the solid in terms of L.

2. The volume of the hole increased exponentially. At midnight, the volume of the hole was 10 cubic meters. Two hours later, the volume of the hole was 30 cubic meters. How large would the hole be 5 hours after midnight?

Integrate:

3. $\int 2 \cos x \, dx$

4. $\int \frac{3\, dt}{\sqrt{t}}$

5. $\int 2\sqrt[3]{u}\, du$

Use implicit differentiation in problems 6–8:

6. Find $\frac{dy}{dx}$ if $2xy - y^2 = \sin x$.

7. Suppose u and v are both functions of time. Find $\frac{du}{dt}$ if $u^3 + v^2 = uv$.

8. Find the slope of the line which can be drawn tangent to the graph of $y^2 + x^2 = 2$ at the point (1, 1).

9. Which of the following graphs most resembles the graph of $y = x^2(x - 2)^3$?

(a)
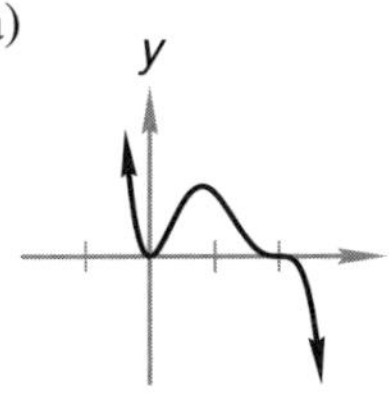

(b)
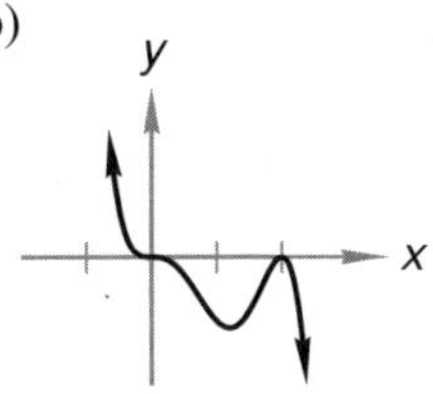

(c)
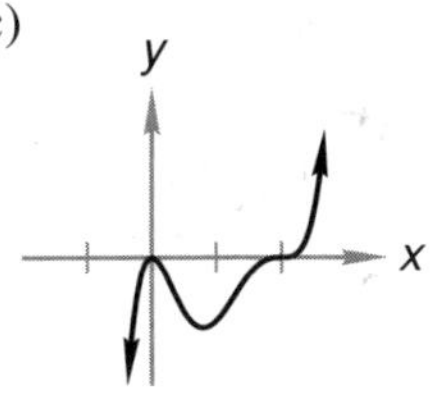

(d)
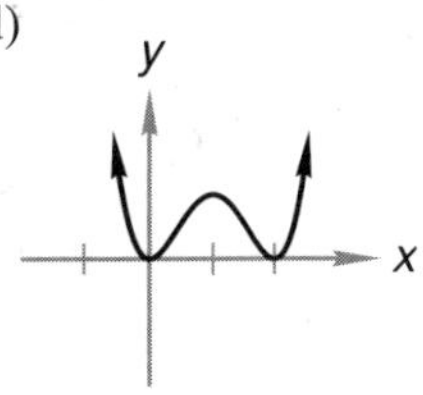

10. Make a rough sketch of all the possible shapes of the graph of $y = x^4 + ax^3 + bx^2 + cx + d$ where a, b, c, and d are real numbers.

11. Find $f''(1)$ if $f(x) = \ln|x| + x^2$.

Evaluate the following limits:

12. $\lim_{h \to 0} \frac{e^{x+h} - e^x}{h}$

13. $\lim_{x \to 1} \frac{x^2 + x - 2}{x - 1}$

14. Sketch the graph of $y = -|x|$ and use interval notation to describe those values of x for which the function is increasing.

15. Describe the domain and range of $y = \sqrt{x - 1}$.

Concept Review:

16. Find the area of an equilateral triangle, each of whose sides has length 6.

1. The volume V (measured in cubic centimeters) of the balloon at a given time t (measured in seconds) is given by the equation

$$V(t) = t^2 + 2t$$

Find the rate of change of the volume at $t = 3$ seconds.

2. A particle moves along the number line so that its directed distance from the origin at a given time t is given by

$$s(t) = -2 \sin t$$

Find the velocity (the rate of change of s) of the particle at $t = \pi$ seconds.

3. Find the slope of the normal line which can be drawn to the graph of $y = 2 \ln |x|$ at $x = -1$.

4. Find the critical numbers of f where $f(x) = \frac{1}{3}x^3 - x$. Use a rough sketch of the function and its equation to find where the maximum and minimum of f occur and what their corresponding values are.

Integrate:

5. $\int \left(2e^x - \cos x + \sqrt{x}\right) dx$

6. $\int \left(\frac{2}{t} + 3\right) dt$

Use u substitution to differentiate the functions in problems 7 and 8:

7. $y = \sin^3 x$

8. $y = \sqrt{x^2 - 5}$

9. Use implicit differentiation to find $\frac{dy}{dx}$ if $\sin x - \cos y = y$.

10. Find $\frac{dA}{dt}$ where $A = 4\pi r^2$ and where the area A and the radius r are both functions of time t.

11. Differentiate as many times as necessary to evaluate $f'''(-3)$ if $f(x) = 2 \ln |x|$.

12. Find $\frac{dy}{dx}$ if $y = e^x \sin x$.

13. Sketch the graph of g where $f(x) = x^2 - 3x + 2$ and $g(x) = \frac{1}{f(x)}$.

Evaluate the following limits:

14. $\lim_{x \to 1} \frac{x^2 + x - 2}{x - 1}$

15. $\lim_{x \to \infty} \frac{4x^3 - x}{1 - 2x^3}$

Concept Review:

16. Find the area of the square which can be circumscribed about a circle of radius 4.

1. The volume of liquid decreased exponentially. At midnight there were 4 liters of liquid, and 2 hours later there were only 3 liters of liquid. How much liquid would there be at 6:00 a.m.?

Use the chain rule to find $\frac{dy}{dx}$ in problems 2 and 3:

2. $y = \cos u$ and $u = 4x^2$

3. $y = \ln |u|$ and $u = \sin x$

4. Use the definition $f'(a) = \lim_{x \to a} \frac{f(x) - f(a)}{x - a}$ to find $f'(2)$ if $f(x) = x^2$.

5. Sketch the graph of $y = 2x + 1$ and partition the interval $[0, 4]$ into four subintervals of equal length. Estimate the area under the graph over the specified interval by computing an upper sum.

Use the quotient rule to differentiate the functions of problems 6 and 7:

6. $f(x) = \frac{e^x}{\sin x + 1}$

7. $g(x) = \frac{x^2 + 1}{x + e^x}$

8. Sketch the graph of $y = \frac{(x^2 + 1)(x - 2)}{(x + 1)^2(x - 3)}$.

9. A particle moves along the number line so that its position at time t is given by $x(t) = 3 \ln |t - 1|$. Find the velocity of the particle at $t = 2$ seconds.

10. Find the equation of the normal line which can be drawn to the graph of $y = x^2 + \ln |x|$ at $x = 1$. Use a calculator as required.

Integrate:

11. $\int \left(\frac{1}{\sqrt{t}} - 4 \sin t + 1 \right) dt$

12. $\int \left(\frac{3}{x} + e^x \right) dx$

13. Find $(fg)'(x)$ if $f(x) = 2e^x$ and $g(x) = \cos x$.

14. If L and x are both functions of time, use implicit differentiation to find $\frac{dL}{dt}$ if $\frac{L - x}{8} = \frac{L}{4}$.

15. Evaluate: $\lim_{x \to \infty} \frac{x^2 + 1}{x + 1}$.

16. On which of the following intervals is $\ln (\sin x)$ defined?

a. $0 < x < \pi$

b. $-\frac{\pi}{2} \le x \le \frac{\pi}{2}$

c. $0 \le x \le \pi$

d. $-\frac{\pi}{2} < x < \frac{\pi}{2}$

Concept Review:

17. Find x in the figure shown.

$m\widehat{ABC} = \left(\frac{1}{2}x + 72 \right)^\circ$

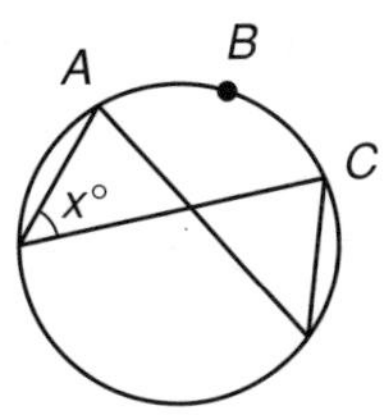

1. A 13-meter-long ladder leans against a vertical wall. The base of the ladder is being pulled away from the wall at a rate of 1 meter per second. How fast is the top of the ladder falling when the base of the ladder is 5 meters away from the wall?

2. Express the cosine function in terms of the sine function and use the fact that $\frac{d}{dx}(\sin x) = \cos x$ to differentiate $y = \cos x$.

Differentiate the following functions:

3. $y = e^x \sec x$

4. $y = (\ln x)(\tan x)$

5. $y = \csc x + \cot x$

Use definite integration to compute the area of the shaded regions shown:

6.

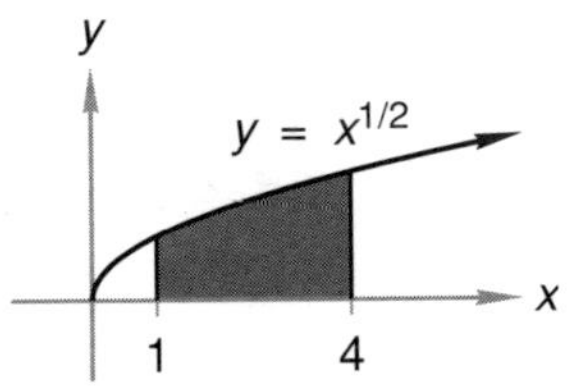

7.

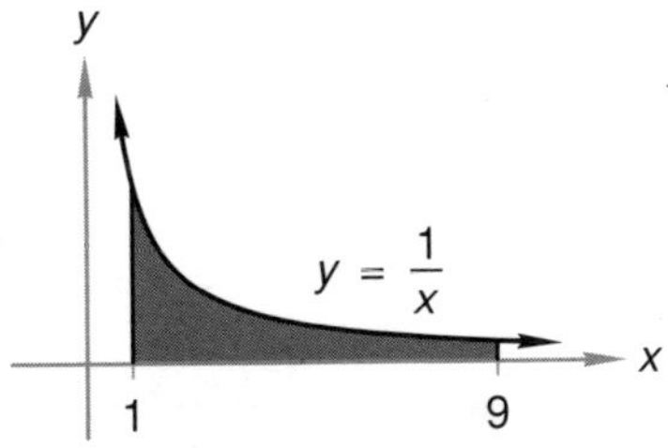

8.

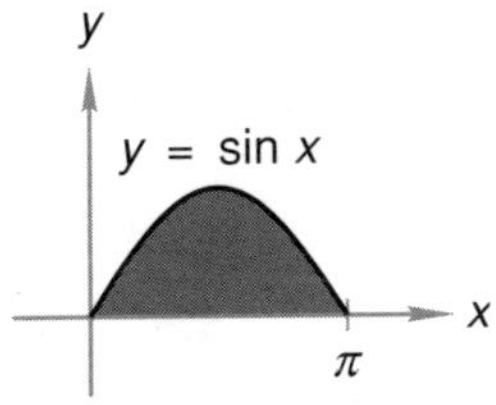

9. Suppose $f(x) = x^2 + 2x - 3$. Use the first derivative to determine where the minimum of f occurs. Use a rough sketch of f to justify your answer.

10. Suppose that g is a function such that $g'(1) = 0$ and $g' > 0$ when x lies in the interval $(-1, 1)$ and $g' < 0$ when x lies in the interval $(1, 3)$. Sketch the shape of g for x values near $x = 1$. Indicate any special characteristics of g at $x = 1$.

11. Use the chain rule to find $\frac{dy}{dx}$ if $y = 2\cos u$ and $u = x^2 + 1$.

12. A particle moves along the number line so that its position at time t is given by $s(t) = t^3 + t^2 + 10$. Find the velocity of the particle (rate of change of position) at $t = 1$.

Integrate:

13. $\int \left(-\frac{1}{u} + u^2 + 1\right) du$

14. $\int \left(\frac{x^2 - x}{x}\right) dx$ (divide first)

Concept Review:

15. Express $\log_2 x$ entirely in terms of the natural logarithm.

1. Use the first derivative and a rough sketch of f to determine the x values of the local maxima and minima of $f(x) = \frac{1}{3}x^3 - \frac{1}{2}x^2 + 1$.

2. Find the coordinates of any points of inflection of the graph of f in problem 1. Use the second derivative test to justify your answer to problem 1.

3. Stig and Cristina wanted to enclose a rectangular plot of land that adjoins a straight brick wall. If they had only 100 yards of fence, what should be the dimensions of the rectangular plot so that the area enclosed is a maximum?

Integrate by guessing a form $u^n\,du$, $\frac{du}{u}$, or $e^u\,du$:

4. $\int 4 \cos x \sin^3 x\, dx$

5. $\int \frac{3x^2}{x^3 + 1}\, dx$

6. $\int 3x^2 e^{x^3}\, dx$

7. Expand and then integrate: $\int (x^2 + 1)^2\, dx$

Differentiate:

8. $y = xe^x$

9. $y = 2 \ln |x^2 + 2x|$

10. $y = \frac{\cos x}{(x^2 + 1)^2}$

11. $y = \cos (x^2 + 1)$

12. Use definite integration to compute the area of the shaded region shown.

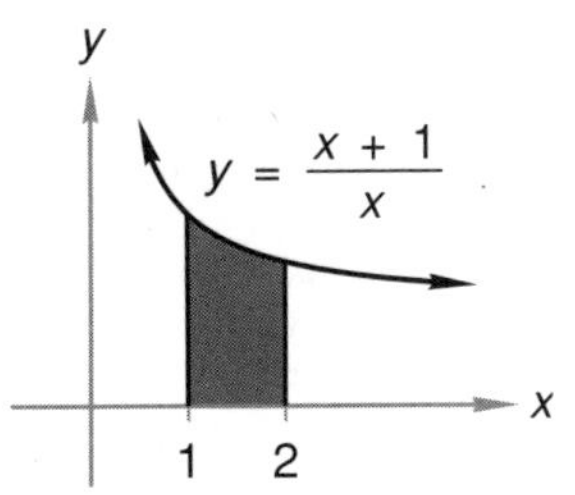

Evaluate the following limits:

13. $\lim_{n\to\infty} \frac{1 - 3n^2}{n^2 + 100}$

14. $\lim_{h\to 0} \frac{e^{x+h} - e^x}{h}$

Concept Review:

15. Determine the domain of the function $f(x) = \frac{\sqrt{x - 1}}{x}$.

1. A line joining the points (0, 8) and (4, 0) is shown. Find the coordinates of the point (x, y) which maximizes the area of the inscribed rectangle.

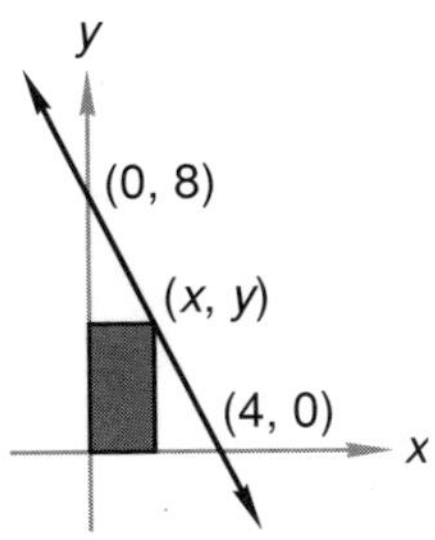

2. A particle moves in a circular orbit described by the equation $x^2 + y^2 = 25$. As it passes through (3, 4), its y coordinate is decreasing at a rate of 1 unit per second. What is the rate of change of the x coordinate the instant the particle is passing through the point (3, 4)?

3. A particle is moving along the number line so that its position at time t is given by:

$$s(t) = t^3 - 12t$$

Find the times $(t > 0)$ when the particle is momentarily at rest.

4. If $\int_{-1}^{3} f(x)\,dx = 2$ and $\int_{3}^{6} f(x)\,dx = -4$, find $\int_{-1}^{6} f(x)\,dx$.

5. If $\int_{1}^{4} f(x)\,dx = 5$ and $\int_{1}^{4} g(x)\,dx = -3$, find $\int_{1}^{4} [2f(x) - 3g(x)]\,dx$.

6. If f is a function which is continuous on $[-1, 4]$ and attains a maximum value of 5 on $(-1, 4)$, then which of the following must be true?

a. $\int_{-1}^{4} f(x)\,dx \le 25$ **b.** $\int_{-1}^{4} f(x)\,dx \le 20$ **c.** $\int_{-1}^{4} f(x)\,dx \ge 0$ **d.** $\int_{-1}^{4} f(x)\,dx \le 5$

Evaluate:

7. $\int_{1}^{4} \frac{1}{\sqrt{x}}\,dx$ **8.** $\int_{-\pi/2}^{\pi} \cos x\,dx$

Integrate:

9. $\int \frac{x + 1}{x^2 + 2x}\,dx$ **10.** $\int (4x - 2)e^{x^2 - x}\,dx$

11. Make a rough sketch of f if $f'(2) = 0$ and if $f''(x) > 0$ when $x < 2$ and if $f''(x) < 0$ when $x > 2$.

12. Find the area of the region between the x axis and the graph of $y = \frac{1}{x}$ over the interval $[1, e^2]$.

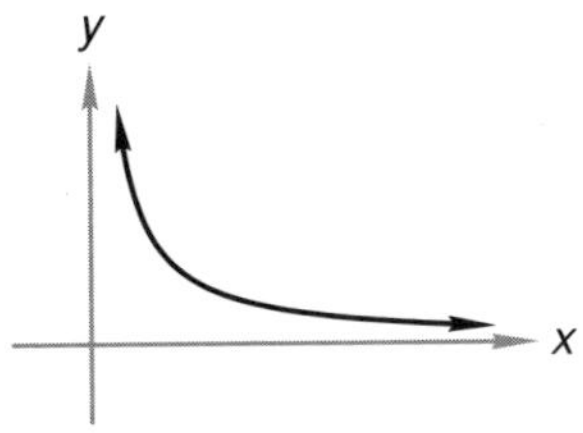

13. Shown is the graph of f'.

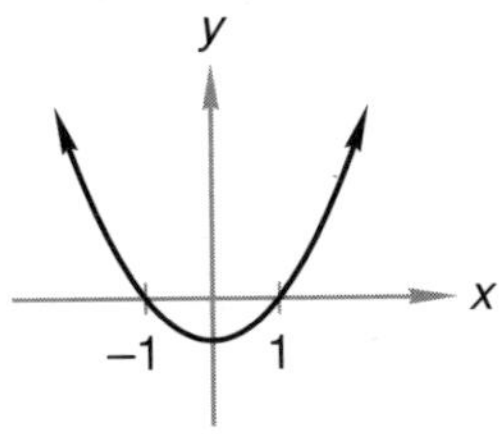

Which of the following could be the graph of f?

(a)

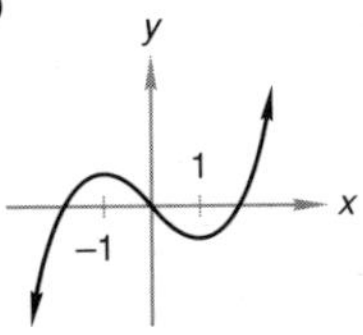

(b)

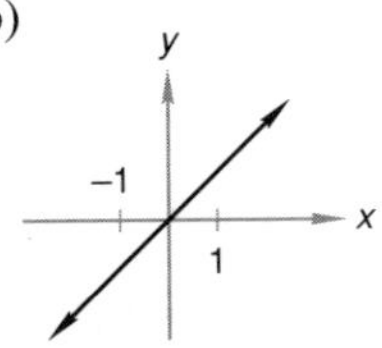

(c)

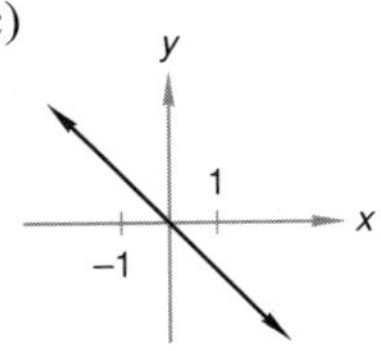

(d)

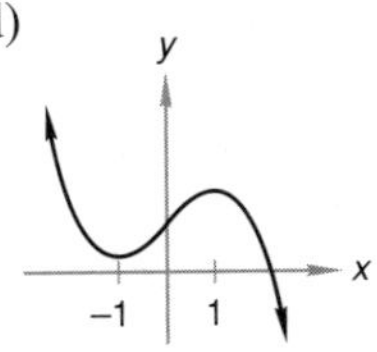

(e)

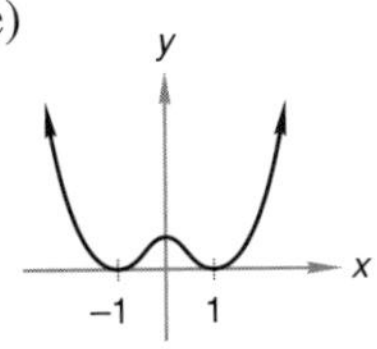

14. Find $\frac{dy}{dx}$ if $y = \frac{e^{x+2} + x}{\sin x - e^x} + \sec 3x$.

Concept Review:

15. Without using a calculator, determine the exact value of sin 20° sec 70°.

1. A hemispherical bowl has a diameter of 12 centimeters. If water is poured into the bowl at a rate of 1 cubic centimeter per second, how fast is the water rising when the water is 3 centimeters deep? The formula for the volume of liquid in a hemisphere of radius r filled to a depth h is: $V = \pi r h^2 - \frac{1}{3}\pi h^3$.

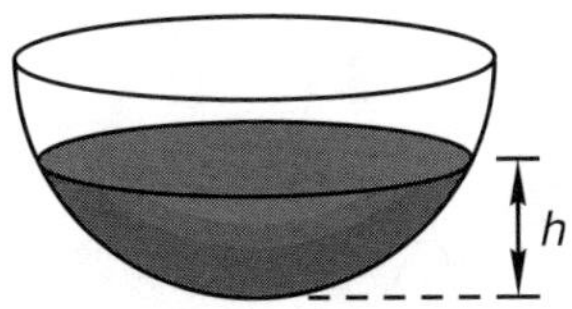

2. The equation of f is $f(x) = ax^3 + bx$. Find a and b if the graph of f passes through the point $(1, 3)$ and the slope of the graph of f at $x = 1$ is 7.

3. Shown is the graph of f''. Sketch a possible graph for f.

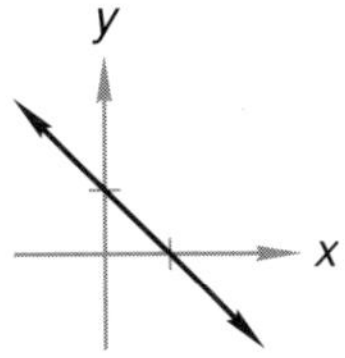

4. If $f''(x) = 4$ for all values of x, $f'(1) = 6$, and $f(1) = 5$, find the equation of f.

Integrate:

5. $\int \cos 3x \, dx$

6. $\int \frac{x^2 + 1}{x} \, dx$

7. Find the area of the region in the first quadrant bounded by the graph of $y = x\sqrt{4 - x^2}$

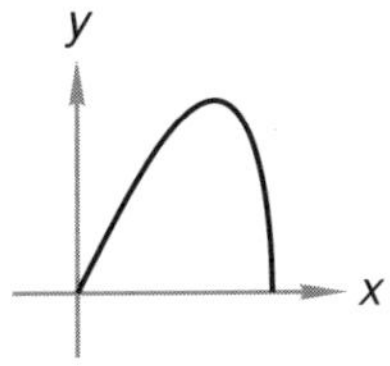

8. Find the area of the region which is completely bounded by the graph of $f(x) = x^2 - 2x$ and the x axis.

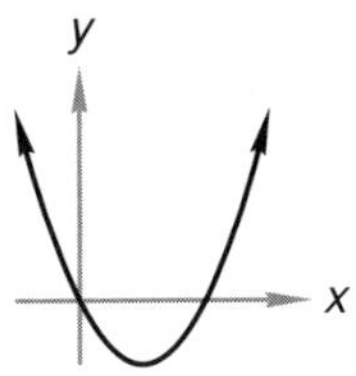

9. Find the area completely enclosed by the graphs of $y = x^2$ and $y = 2x^2 - 1$.

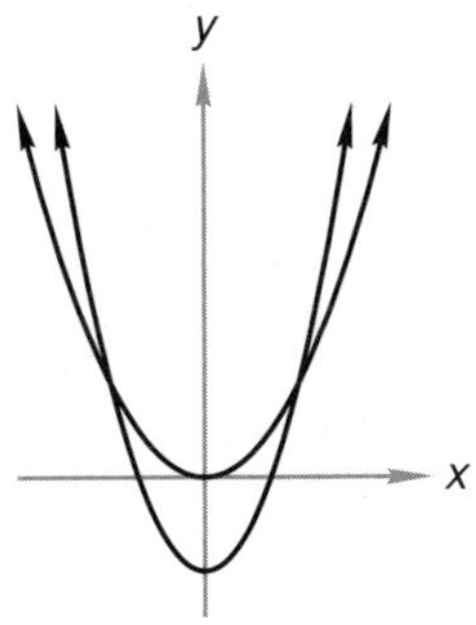

10. Find k if $\int_{-1}^{1} (4x^3 + k) \, dx = 10$.

11. Sketch the graph of $y = \frac{(x - 2)(x - 4)}{x(x + 1)(x - 3)(x - 4)}$.

12. Find $f^{-1}(2)$ if $f(x) = \ln x$, where f^{-1} denotes the inverse of f.

13. Write the equation of the inverse of $y = \sin x$ where $-\frac{\pi}{2} \le x \le \frac{\pi}{2}$. The variable y may be expressed either explicitly or implicitly as a function of x.

14. Differentiate: $y = \frac{1}{\sqrt{x}} + \sqrt{x} + e^{\sin x} + \cos 2x$.

Concept Review:

15. Write the explicit equation for the part of the graph of the ellipse $\frac{x^2}{a^2} + \frac{y^2}{b^2} = 1 \quad (a, b > 0)$ which lies in the first two quadrants.

1. Suppose $\tan\theta = \dfrac{x}{100}$ and θ and x are functions of time. Find $\dfrac{d\theta}{dt}$ when $\sec\theta = \dfrac{5}{2}$ and $\dfrac{dx}{dt} = -200$. (Begin by differentiating implicitly with respect to time.)

2. A ball is dropped from a height of 200 meters. Write equations which express the height, velocity, and acceleration of the ball as a function of t, where t is the time in seconds after the ball is released.

3. Begin with the equation $f(x) = y = \cos x$, $0 \le x \le \pi$. Write an equation which expresses the inverse of f implicitly. Differentiate this equation implicitly and use an appropriately labeled right triangle to find the equation for $\dfrac{d}{dx}f^{-1}$.

Differentiate:

4. $y = \arcsin\dfrac{x}{2}$

5. $y = \arctan\dfrac{x}{3}$

6. Apply the critical number theorem to find where the maximum and minimum values of $f(x) = x^2 + 1$ occur on the interval $[-2, 1]$. What are the maximum and minimum values of f on this interval?

7. A function f is continuous on the closed interval $[-2, 3]$. In addition, $f(-2) = 4$, $f(1) = 1$, $f(3) = 10$, and f has the properties listed to the right. Sketch a possible graph of f and determine the maximum and minimum value of f on the interval described.

x	$-2 < x < 1$	$x = 1$	$1 < x < 3$
$f'(x)$	negative	zero	positive

8. A force is applied to an object so that the force at a given position x along a horizontal axis is given by $F(x) = 3x^2$ newtons. Find the work done to the object if it is moved from $x = 2$ meters to $x = 3$ meters.

9. Suppose $f(x) = ax^2 + bx + c$. Find a, b, and c if $f(1) = 2$, $f'(1) = 5$, and $f''(1) = 4$.

Integrate:

10. $\displaystyle\int \frac{\cos x}{\sqrt{1 + \sin x}}\,dx$

11. $\displaystyle\int xe^{x^2+1}\,dx$

12. Find the area completely enclosed by the graphs of $y = x$ and $y = x^2$.

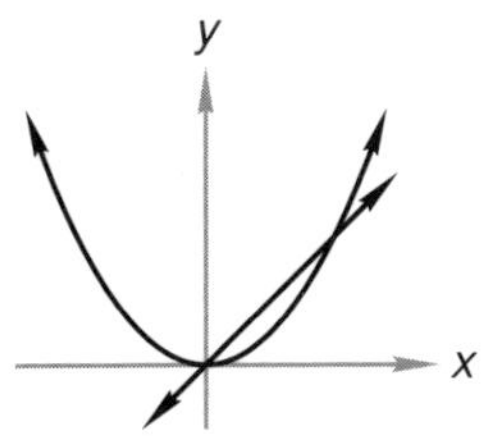

13. Suppose $2 \le c \le 6$ and f is continuous on $[2, 6]$. Find $\displaystyle\int_2^c f(x)\,dx$ if $\displaystyle\int_2^6 f(x)\,dx = 12$ and $\displaystyle\int_c^6 f(x)\,dx = -4$.

14. Differentiate: $y = \ln\left|x^2 + 1\right| + \dfrac{e^{2x}}{\sec x + 1}$.

Concept Review:

15. f is a function such that $f(xy) = f(x) + f(y)$. Which of the following could be the equation of f?

a. $f(x) = \log x$

b. $f(x) = e^x$

c. $f(x) = x^2$

d. $f(x) = 3x$

1. A rectangle whose sides are changing in length has a constant area of 100 square meters. Find the length of the rectangle when its width is decreasing at a rate of 1 meter per second and its length is increasing at a rate of 1 meter per second.

Use integration by parts to compute the following integrals:

2. $\int 2xe^x\,dx$

3. $\int x\sin x\,dx$

4. Determine whether the following function is odd or even, if either: $y = \dfrac{\sin x}{x^2}$.

5. Is the graph of $y = x + \sin x$ symmetric about the y axis, the origin, or neither?

Differentiate to solve problems 6 and 7:

6. $\dfrac{d}{dx}\dfrac{e^{\sin 2x}}{x^2 + 1}$

7. If $y = \arcsin e^x$, find y'.

8. Integrate: $\int xe^{x^2}\,dx + \int \dfrac{1}{\sqrt{x}}\,dx$

9. Which of the following definite integrals equals $\int_0^{\pi/4} (\cos 2x)e^{\sin 2x}\,dx$?

a. $\int_0^1 e^u\,du$ **b.** $\int_0^1 -e^u\,du$ **c.** $\frac{1}{2}\int_0^1 e^u\,du$ **d.** $\frac{1}{2}\int_0^{\pi/4} e^u\,du$

10. Find the area of the region completely enclosed by the graphs of $y = x^2 - 2$ and $y = x$.

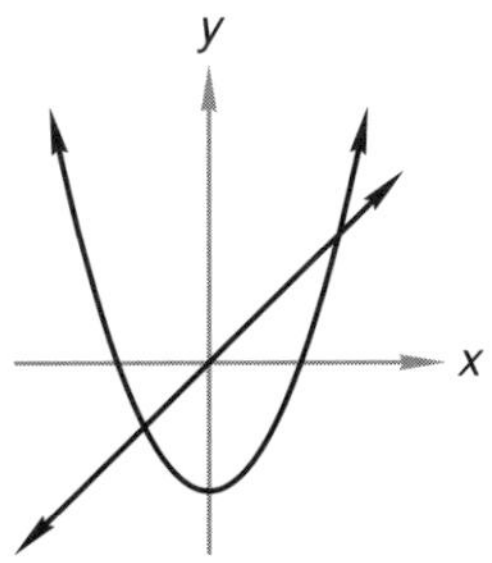

11. Write the definite integral in the variable y which equals the area of the region completely enclosed by the graphs $x = y^2 - 2$ and $y = -x$.

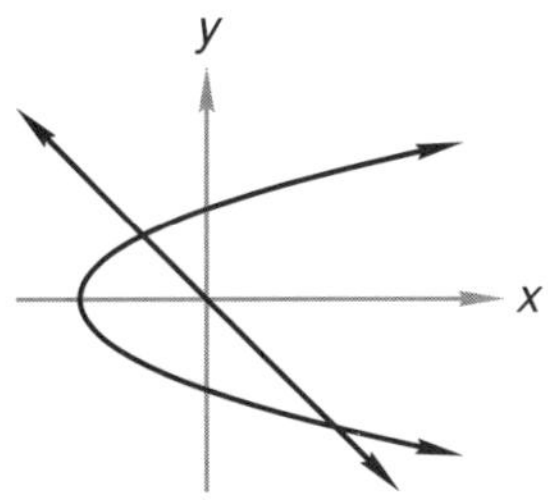

12. Suppose $f(x) = 3x + 1$. Find the equation of f^{-1}. Evaluate $f^{-1}(2)$ and $(f^{-1} \circ f)(x)$.

13. If $\int_{-1}^{5} f(x)\,dx = 10$ and $\int_{-1}^{1} f(x)\,dx = -3$, find $\int_1^5 f(x)\,dx$.

14. Find the slope of the tangent line which can be drawn to the graph of $y = \ln x$ at $x = 2$.

Concept Review:

15. Determine which of the following equations describes a curve which satisfies the following property: For every point (x, y) which lines on the curve, $(-x, -y)$ also lies on the curve.

a. $y = e^x$ **b.** $x^2 + y^2 = 4$ **c.** $y = 3x + 1$ **d.** $x^3 + y^3 = 1$

1. Boyle's law tells us that if the temperature of a quantity of ideal gas is unchanged, the product of the pressure and the volume equals a constant k. Suppose we have 100 m^3 of gas at a pressure of 20 N/m^2, and the volume is decreasing at a rate of 1 m^3 per second. Find the rate at which the pressure is changing if the volume is 50 m^3.

2. A 4-m-deep tank is completely filled with a fluid whose weight density is 1000 N/m^3. Find the total force exerted on one end of the tank if the end of the tank is rectangular as shown.

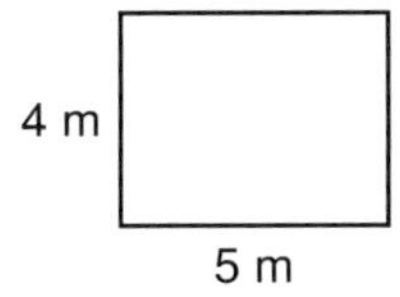

3. A container with a triangular end shown is filled with a fluid whose weight density is 6000 N/m^3. Find the total force against the end of the container shown.

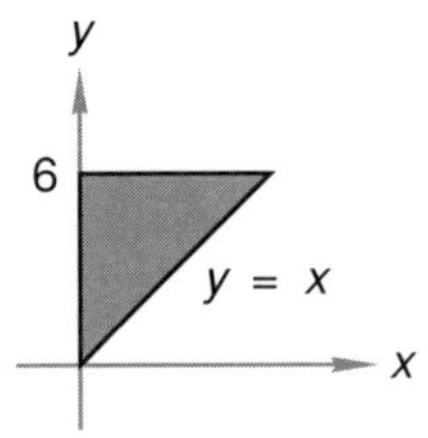

4. Let R be the region bounded by the graph of $y = 1 - x$ and the coordinate axes. Use the "disk method" to write a definite integral whose value equals the volume of the solid formed when R is revolved about the x axis. Evaluate this definite integral.

5. Let R be the region between the graph of $y = x^2$ and the x axis from $x = 0$ to $x = 1$. Use the disk method to write a definite integral whose value equals the volume of the solid formed when R is revolved about the x axis. Evaluate this definite integral.

Differentiate:

6. $y = \log_6 x - \log_4 x + 3 \log_3 x$
(Use your calculator to simplify answer.)

7. $y = (2.7)^x + 4^x + e^{5x}$

Integrate:

8. $\int \log_4 x \, dx$

9. $\int 2xe^x \, dx$

10. $\int 2xe^{x^2} \, dx$

11. $\int \frac{x}{\sqrt{x^2 - \pi}} \, dx$

Differentiate:

12. $y = \arcsin 2x + \frac{e^x}{x + \sin x}$

13. Which of the following limits is undefined?

a. $\lim_{x \to 2} \frac{x^3 - 8}{x - 2}$ **b.** $\lim_{x \to 0^-} \frac{|x|}{x}$ **c.** $\lim_{x \to 0^+} x \sin x$ **d.** $\lim_{x \to 0} \sin\left(\frac{2}{x}\right)$

14. Find the slope of the tangent line which can be drawn to the function $xy = 1$ at $x = 1$.

Concept Review:

15. Suppose f is a function which is defined for all real numbers. Which of the following conditions guarantees that the inverse of f is also a function?

a. f is a strictly decreasing function
b. f is an odd function
c. f is a periodic function
d. f is continuous and differentiable everywhere
e. Any vertical line intersects the graph of f at most at one point

1. Shown is the graph of $y = 2e^{-x^2}$ and a rectangle, two of whose corners touch the graph. Find the value of x for which the area of the rectangle is a maximum.

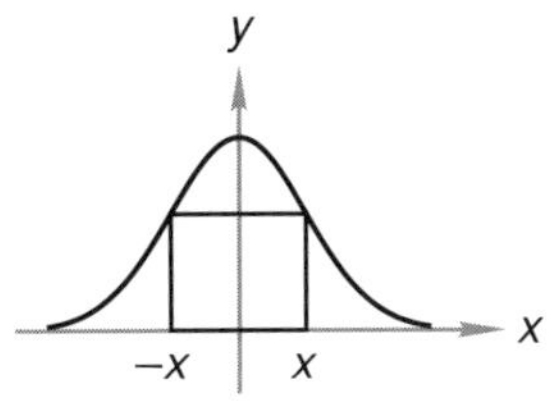

2. A particle moves along the x axis so that its acceleration is given by

$$a(t) = 6t - 2$$

If its velocity at $t = 1$ is 2 and its position at $t = 0$ is 2, develop the equations which express the particle's position and velocity as a function of t.

3. A 6-meter-long tank with a semicircular cross-sectional diameter of 6 meters, as shown, is partially filled with a fluid whose weight density is 3000 newtons per cubic meter. If the depth of the water in the tank is 2 meters, write a definite integral which expresses the work done in completely pumping the fluid out of the tank.

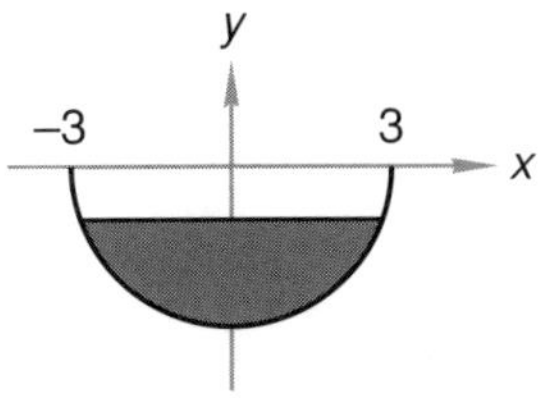

$x^2 + y^2 = 9$

4. Use a Pythagorean identity to aid in finding $\int \sin^2 x \cos^3 x\, dx$.

5. Is the following statement true or false? Given that f is a function such that $\lim_{x\to 1^-} f(x) = \lim_{x\to 1^+} f(x)$, f is continuous at $x = 1$. Explain why.

6. Approximate to two decimal places the area between the graph of $y = 3^x$ and the x axis from $x = 1$ to $x = 3$.

Integrate:

7. $\int x \cos x\, dx$

8. $\int \frac{x - 1}{\sqrt{x}}\, dx$

9. $\int \frac{x}{x^2 - 1}\, dx$

10. $\int (\sin x)(\cos x - \pi)^3\, dx$

11. Use y as the variable of integration to write a definite integral whose value equals the area of the region in the first quadrant bounded by the graphs of $y = 2x^2$ and $y = 1$.

12. Suppose f is a function continuous on the interval $[-1, 2]$, $f(-1) = 4$, $f\left(-\frac{1}{2}\right) = 6$, $f(0) = 3$, and $f(2) = 1$ and the functions f' and f'' have the properties described in the table shown.

	$-1 < x < -\frac{1}{2}$	$x = -\frac{1}{2}$	$-\frac{1}{2} < x < 0$	$x = 0$	$0 < x < 2$
f'	positive	zero	negative	negative	negative
f''	negative	negative	negative	zero	positive

Sketch f and list where f attains its absolute maximum and minimum on the interval given, and tell what their values are. Also, give the coordinates of any points of inflection.

13. Find $y = \frac{d}{dx} \arcsin 3x + \int \frac{3}{\sqrt{1 - 9x^2}}\, dx$.

14. Suppose f is a function continuous on the closed interval $[0, 2]$ and $2 \le f(x) \le 6$. What is the greatest **possible** value that $\int_0^2 f(x)\, dx$ could have?

a. 0 **b.** 10 **c.** 12 **d.** 15

Concept Review:

15. Sketch the graph of $y = \ln x^2$, $x > 0$. Begin by rewriting the equation of the function in a simpler form by using a property of logarithms.

1. A 6-meter-long trough with a right triangular cross section as shown is partially filled with a fluid of weight density 9000 newtons per cubic meter. If the level of the fluid is 1 meter below the top rim of the trough, find the work done in pumping all the fluid out of the tank.

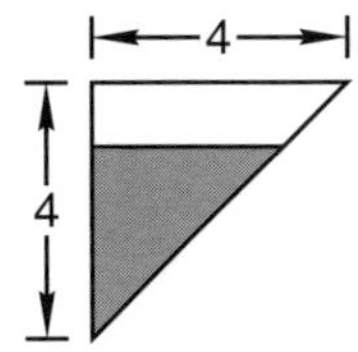

2. Suppose the function f is defined as follows: $f(x) = \begin{cases} x^2 + 2x & \text{when } x \le 2 \\ x + b & \text{when } x > 2 \end{cases}$

   Find the value of b which makes f continuous everywhere.

3. Let f be a quadratic function. The slope of the tangent line to the graph of f at $x = 1$ is 3, and the slope of the tangent line which can be drawn to the graph of f at $x = 2$ is 5. If the graph of f passes through the point (0, 1), find the equation of f.

4. Let R be the region completely bounded by the graph of $y = \sqrt{x}$ and $y = x$. Find the volume of the solid formed when the region R is revolved about the x axis.

5. Let R be the region completely bounded by the graphs of $y = x$, $y = \frac{1}{2}x$, and $y = 1$. Use the "washer method" to write a definite integral whose value equals the volume of the solid formed when the region R is rotated about the y axis. Evaluate this definite integral.

6. Find the balance point for this system of masses.

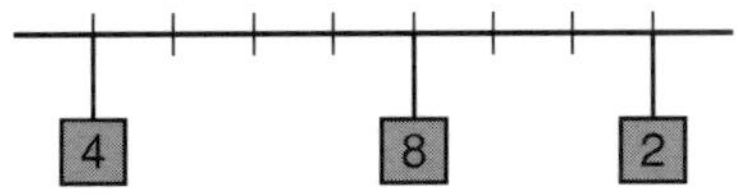

7. Make a rough sketch of $y = \dfrac{(x - 2)(x + 1)}{x}$.

8. Write the equations of all the asymptotes of the graph of the function $y = \dfrac{x}{x^3 + 4x^2 + x - 6}$.

Evaluate the following limits:

9. $\displaystyle\lim_{x \to \pi} \frac{\sin (x - \pi)}{x - \pi}$

10. $\displaystyle\lim_{x \to \infty} \frac{x + 1}{e^x}$

11. Find the slope of the line which can be drawn normal to the graph of $y = \arcsin \dfrac{x}{2}$ at $x = -1$.

12. Differentiate: $y = x \ln |x^3 - 1| + 2^{x-1} + \arctan x$.

13. Integrate: $\displaystyle\int \left(3^x + \frac{1}{x^2 + 1}\right) dx$

14. Evaluate: $\displaystyle\int_0^{\pi} \frac{\cos x}{\sqrt{\sin x + 1}}\, dx$ by using u substitution.

Concept Review:

15. If the graph of f is as shown at right, which of the following graphs most resembles the graph of g where $g(x) = f(x + 2)$?

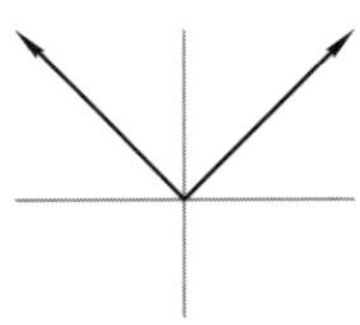

(a)

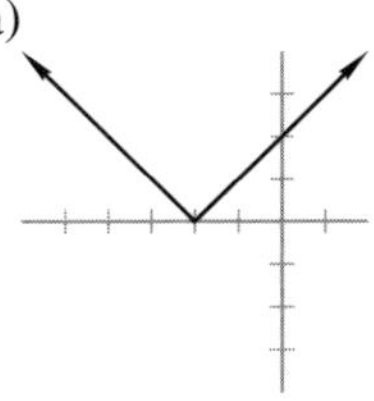

(b)

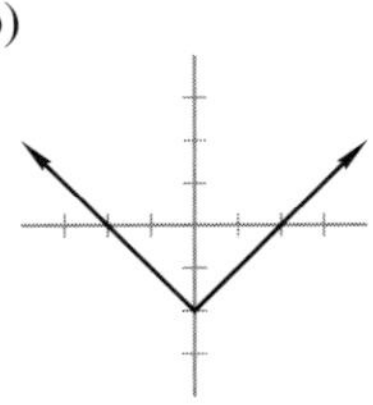

(c)

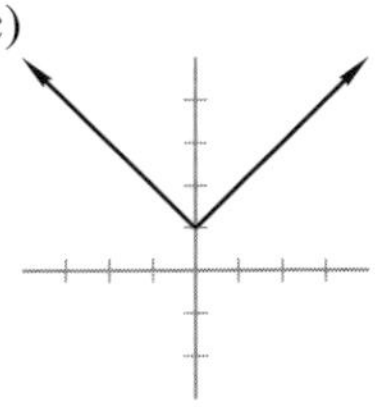

(d)

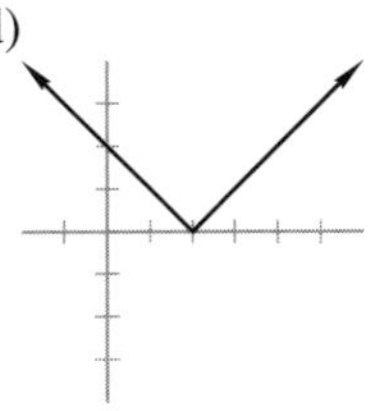

(e)

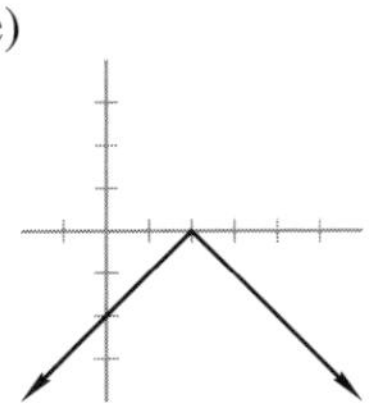

1. A spring with spring constant $k = 3$ newtons per meter is stretched from $x = 1$ meter to $x = 4$ meters. Find the work done to the spring.

2. A ball is thrown horizontally from the top of a 200-meter-tall building. The height of the ball above the ground at time t after the ball has been thrown is exactly the same as if the ball had been dropped. Find the equation which describes the height of the ball as a function of t.

3. Approximate to two decimal places the time t, after the ball of problem 2 is thrown, that it will take to strike the ground. If the ball is thrown with a horizontal velocity of 20 meters per second, what is the horizontal distance the ball will travel?

4. A cylindrical container 2 meters high with a circular base of radius 1 meter is completely filled with a fluid of weight density 100 newtons per cubic meter. Find the work performed in pumping all the fluid out of the cylinder.

5. Let f be a function defined as follows:

$$f(x) = \begin{cases} x^3 & \text{when } x < 1 \\ ax + b & \text{when } x \geq 1 \end{cases}$$

Determine the values of a and b that will make f both continuous and differentiable everywhere.

Use logarithmic differentiation to differentiate the following functions:

6. $y = \dfrac{x^3\sqrt{x^2 + 1}}{(x + 1)^3}$

7. $y = x^{2x}$

8. Let R be the region between the graph of $y = e^{-x}$ and the x axis on the interval $[1, 2]$. Express the x coordinate of the centroid of R as the quotient of two integrals.

9. Integrate: $\int 3 \sin^2 x\, dx$

10. Evaluate: $\lim_{x \to 0} \dfrac{2 \sin 2x}{5x}$

11. Determine the slope of the tangent line which can be drawn to the graph of $y = \log_2 x$ at $x = 1$.

12. Determine the area between the graph of $y = xe^{-x^2}$ and the x axis over the interval $[0, 1]$. (The function is nonnegative over this interval.)

13. Differentiate: $y = \ln\left|x^2 + 1\right| + \dfrac{x + 1}{e^x + 1}$

14. Integrate: $\int \cos x\sqrt{1 + \sin x}\, dx$

Concept Review:

15. The function $y = x^3 + x$ has
   - **a.** no inflection points
   - **b.** one inflection point
   - **c.** two inflection points
   - **d.** exactly two local maximum points.

1. If money is compounded continuously, the rate of increase is proportional to the money present, B. This statement can be expressed as a differential equation: $dB/dt = kB$. The solution to this differential equation is $B = Pe^{rt}$, where B is the balance in the account t years after the initial deposit, P is the amount of the initial deposit, and r is the annual interest rate. Suppose the annual interest rate is 7 percent. How much money should be invested now at 7 percent for the account to contain \$20,000 after 20 years?

2. The slope of the graph of a function at a given point is 4 times the value of the x coordinate of that point. If the graph of f passes through (1, 3), find the equation of f.

3. The region R is bounded by the x axis and the graph of $y = e^x$ on the interval [0, 1]. Use the "shell method" to write a definite integral whose value is the volume of the solid formed when R is revolved about the y axis. Evaluate this definite integral.

4. Let R be the region bounded by the x axis and the graph of $y = \tan x$ on the closed interval $\left[0, \frac{\pi}{4}\right]$. Use the "disk method" to find the volume of the solid formed when R is revolved about the x axis. Evaluate this integral.

5. Find a general solution to the differential equation $x\,dx + 2y\,dy = 0$.

6. Find a general solution to the differential equation $\frac{dy}{dx} = -x^2y^3$.

7. Suppose $f(x) = \sin x$ and $g(x) = x^5 + x^3$. Determine whether fg is odd, even, or neither.

8. Evaluate: $\int_{-k}^{k} (fg)(x)\,dx$ if $\int_{0}^{k} (fg)(x)\,dx = 5$ and if f and g are as defined in problem 7.

9. Find the area of the region between $y = \sin^3 x$ and the x axis on the interval $[0, \pi]$. (*Note:* $y = \sin^3 x$ is nonnegative on this interval.)

Evaluate the following limits:

10. $\lim_{x\to 1} \frac{x^3 - 1}{x - 1}$

11. $\lim_{x\to 0^+} \frac{2|x|}{x}$

12. $\lim_{h\to 0} \frac{\ln(x + h) - \ln x}{h}$

13. Differentiate: $y = x \sec x + \frac{\sin x}{x^2 + 1} + \arcsin\left(\frac{x}{a}\right)$

14. Integrate: $\int \frac{1}{\sqrt{a^2 - x^2}}\,dx + \int \frac{x}{\sqrt{4 - x^2}}\,dx$

Concept Review:

15. Suppose f is defined as follows:

$$f(x) = \begin{cases} ae^x + \cos x & \text{when } x \geq 0 \\ bx & \text{when } x < 0 \end{cases}$$

Find the values of a and b which make f continuous and differentiable everywhere.

1. A right circular cone is inscribed inside a hemisphere so that its base is the same as the base of the hemisphere. If the radius of the hemisphere is increasing at a rate of 1 cm/sec, find the rate at which the total volume of the cone is increasing when the radius of the base is 4 cm.

2. A rectangular tank 5 meters wide, 6 meters long, and 4 meters deep is filled with a fluid whose weight density is 6000 newtons per cubic meter. Find the work done in completely pumping out the tank.

3. Let R be the region bounded by the graph of $y = x^2$ and the x axis on the interval $[0, 1]$. Find the volume of the solid formed when R is revolved about the line $x = -1$.

4. Let R be the region bounded by the y axis and the graph of $x = 1 - y^2$. Use y as the variable of integration to write a definite integral whose value equals the volume of the solid formed when R is revolved about the line $x = -1$.

5. Let R be the region bounded by the graph of $y = \sin x$ and the x axis on the interval $[0, \pi]$. Write a definite integral whose value equals the volume of the solid formed when R is revolved about the y axis.

6. The function $f(x) = x^3 + x$ is a one-to-one function, and thus its inverse f^{-1} is also a function. Find the equation of the tangent line which can be drawn to the graph of f^{-1} at the point $(2, 1)$.

7. If $f(x) = \tan x$ where $-\frac{\pi}{2} < x < \frac{\pi}{2}$, write the equation of f^{-1} and evaluate $\left(f^{-1}\right)'(1)$.

8. If $f(x) = x^3 + 2$ and $I = [0, 2]$, find a value of c in I for which $f(c)$ equals the average value of f on I.

9. Suppose $f(x) = x^2 + 2$. Find the slope of the line which passes through the points $(0, f(0))$ and $(2, f(2))$. Illustrate the mean value theorem for derivatives by finding a number c in $(0, 2)$ such that $f'(c)$ equals the slope of the line described.

10. Suppose $f(x) = e^x$ and $g(x) = \cos x$ and $h(x) = (f \circ g)(x)$. Find the equation of h' and evaluate $h'\left(\frac{\pi}{2}\right)$.

11. Sketch the graph of $y = \frac{x^2 - x}{x}$.

12. Find $\frac{d}{dx}\left(\cos\left(x^2 + 1\right) + \frac{\cos x - 1}{e^x + 2}\right)$.

13. Find $\int \frac{3x^2}{x^3 + 1}\,dx$

Concept Review:

14. Suppose f is a function and the inverse of f, f^{-1}, is also a function. Which of the following sets of points could lie on f?

a. $\{(1, 3), (2, 1), (3, 2), (4, 4)\}$
b. $\{(1, 4), (4, 1), (4, 4), (2, 6)\}$
c. $\{(3, 5), (1, 4), (-4, 5), (2, 8)\}$
d. $\{(1, 2), (3, 8), (-1, 2), (4, 8)\}$

Test 27, Lesson 106, Form A **SHOW YOUR WORK** **Name:** ____________________

1. Let f be the function $f(x) = xe^{-x^2}$ where $x \in \mathbb{R}$. Find the critical numbers of f. (It is helpful to note that e^{-x^2} is nonzero for any value of x.) Find the intervals on which f is increasing and on which f is decreasing. Use the information found to tell where f attains local maximum and minimum values.

2. A metal ball 200 centimeters in diameter is plated with 0.01 centimeter of gold. Use the method of differentials to estimate the increase in the volume of the ball.

3. A particle moves along the x axis so that its velocity at time t, $t > 0$, is given by $v(t) = \dfrac{t-2}{t}$. If the particle's position at $t = 1$ is $x = 10$, write an equation which describes the position of the particle as a function of t.

4. What velocity does the particle of problem 3 approach as t gets large?

Use the second fundamental theorem of integral calculus to simplify the following expressions.

5. $\dfrac{d}{dx}\displaystyle\int_1^x e^{-1/t}\,dt$

6. $\dfrac{d}{dx}\displaystyle\int_x^5 e^{t^2}\,dt$

Integrate:

7. $\int 2\sin^2 x\,dx$

8. $\int 6\tan^2 x\,dx$

9. $\int 3^x\,dx$

10. $\int \dfrac{2}{\sqrt{1-x}}\,dx$

11. $\int xe^x\,dx$

12. Suppose $f(x) = x^3 + 2x$ and f^{-1} is the inverse of f. Evaluate $f^{-1}(3)$ and $\left(f^{-1}\right)'(3)$.

13. Let R be the region bounded by the graph of $y = e^{x^2}$ and the x axis on the interval $[0, 1]$. Find the volume of the solid formed when R is revolved about the y axis.

14. Illustrate the mean value theorem for derivatives by finding a number c in $[1, 3]$ such that $f'(c) = \dfrac{f(3) - f(1)}{3 - 1}$ where $f(x) = 3x^2 + 4x - 1$.

15. Differentiate: $y = xe^{x^2} - \arcsin(\cos x)$.

Concept Review:

16. If f is a function such that $f' < 0$ for all real values of x, which of the following statements must be true?

a. $f(x) < 0$ for all values of x

b. $f(x_1) < f(x_2)$ for every x_1 and x_2 where $x_1 > x_2$

c. The graph of f is concave upward everywhere

d. The graph of f is concave downward everywhere

1. The height of a right circular cone is increasing at a rate of 6 cm/sec, and the radius of the circular base is decreasing at a rate of 1 cm/sec. Find the rate at which the volume of the cone is changing when the height of the cone is 12 cm and the radius of the base is 3 cm.

2. Use the method of differentials to approximate the cube root of 28.

3. Determine the average value of $f(x) = \frac{1}{x}$ on the interval $I = [1, e]$. Illustrate the mean value theorem for integrals by finding a number c in I where f attains its average value on I.

4. Find the x coordinate of the centroid of the region bounded by the graph of $y = \sin x$ and the x axis on the interval $\left[-\frac{\pi}{2}, 0\right]$.

5. A variable force F is applied to move an object along a straight line such that $F(x) = x^2e^{x^3}$ newtons. Find the work done by the force on the object in moving it from $x = 0$ to $x = 1$ meter.

Use the method of partial fractions to find the following integrals:

6. $\int \frac{x + 2}{(x - 1)(x - 2)}\, dx$

7. $\int \frac{3x + 1}{(x - 2)(x + 1)}\, dx$

8. Write the polar form of $x^2 + y^2 = 9$.

9. Write the rectangular form of the polar equation $r = 3$.

10. Graph $r = \sin 2\theta$ on the polar coordinate plane.

Integrate by using trigonometric substitution:

11. $\int \frac{dx}{\sqrt{1 - 9x^2}}$

12. $\int \frac{dx}{1 + 9x^2}$

13. Evaluate: $\lim_{x \to 0} 5x \csc 2x$

14. Simplify: $\frac{d}{dx}\int_3^x e^{\cot t}\, dt$

Concept Review:

15. If f is a continuous function for all real values of x, which of the following integrals are equal?

(1) $\int_a^b f(x)\, dx$ (2) $\int_{a-1}^{b-1} f(x + 1)\, dx$ (3) $\int_{a+1}^{b+1} f(x - 1)\, dx$

a. (1) and (2) only
b. (2) and (3) only
c. (1) and (3) only
d. (1), (2), and (3)

1. The acceleration for a particle moving along the x axis is $a(t) = 4e^{4t}$. The velocity when $t = 0$ is 5, and the position of the particle when $t = 0$ is 2. Develop the velocity function and the position function of the particle. Find the total distance the particle travels during the time interval [0, 5].

2. Let f be the function defined on the closed interval $I = [-8, 8]$ by $f(x) = \frac{3}{2}x^{2/3} + 3x^{1/3}$.
   Find f' and write the equation of f' as an expression with a single denominator. Determine the critical numbers of f in I. Find at which critical numbers the tangent to the graph of f is horizontal and at which critical numbers the tangent to the graph of f is vertical.

3. Let f be as defined in problem 2. Find x and y coordinates of the absolute maximum and minimum points of the graph of f on the interval defined.

4. Let f be as defined in problem 2. Determine the values of x where the second derivative of f is undefined. Find the x coordinates of all points of inflection of the graph of f. Determine the intervals on which f is concave upward and the intervals on which f is concave downward.

5. Differentiate implicitly to find $\frac{dy}{dx}$ and $\frac{d^2y}{dx^2}$ for the following equation: $y^2 + y = x$.

Integrate by using integration by parts as many times as necessary:

6. $\int x^2 \cos x \, dx$

7. $\int e^x \cos x \, dx$

Integrate by using the method of partial fractions:

8. $\int \frac{4x^2 - 4x - 4}{x^2(x - 2)} \, dx$

9. $\int \frac{-x + 1}{(x + 1)(x + 2)} \, dx$

10. Write the equation of the hyperbola $y^2 - x^2 = 4$ in polar form.

11. Graph $r = 1 + \sin \theta$ on the polar coordinate plane.

12. Evaluate: $\lim_{x \to 0} \frac{3 \cos x - 3}{2 \sin x}$.

13. If $f(x) = \frac{d}{dx} \int_a^x \frac{\cos t + 1}{t} \, dt$, find $f(\pi)$.

14. Evaluate: $\int_{-2}^{3} |x - 1| \, dx$

Concept Review:

15. Let R be the region bounded by the y axis and the graph of $x = 4 - y^2$ on the interval $0 \le y \le 2$. Use y as the variable of integration to write a definite integral whose value equals the volume of the solid formed when R is revolved about the line $x = -1$.